CONTROL TOWER & WEATHER

RAJU SADANAND

INDIA • SINGAPORE • MALAYSIA

Notion Press

Old No. 38, New No. 6
McNichols Road, Chetpet
Chennai - 600 031

First Published by Notion Press 2020
Copyright © Raju Sadanand 2020
All Rights Reserved.

ISBN 978-1-64783-516-3

This book has been published with all efforts taken to make the material error-free after the consent of the author. However, the author and the publisher do not assume and hereby disclaim any liability to any party for any loss, damage, or disruption caused by errors or omissions, whether such errors or omissions result from negligence, accident, or any other cause.

While every effort has been made to avoid any mistake or omission, this publication is being sold on the condition and understanding that neither the author nor the publishers or printers would be liable in any manner to any person by reason of any mistake or omission in this publication or for any action taken or omitted to be taken or advice rendered or accepted on the basis of this work. For any defect in printing or binding the publishers will be liable only to replace the defective copy by another copy of this work then available.

CONTENTS

Foreword 5

Chapter 1: History of Aviation 7

Chapter 2: Navigation 13

Chapter 3: Radio Communications 17

Chapter 4: Ats–Separations 21

Chapter 5: Training 29

Chapter 6: Modernization of Air Traffic Control Establishment 31

Chapter 7: Role of Fire Services in Aviation 33

Chapter 8: Meteorology 37

Chapter 9: Pressure 41

Chapter 10: Temperature 45

Chapter 11: Types of Clouds 47

Chapter 12: Wind 55

Chapter 13: Indian Climatology 59

Conclusion 67

FOREWORD

I am RAJU SADANAND, associated in the Indian Aviation field for the past thirty seven years from 1982. Initially I joined the Indian Meteorological Dept from 1982 and continued upto 1988 then I joined Airports Authority of India as Air traffic controller and continued upto my retirement till 2019 as Deputy General Manager (ATM).

Aviation and its development as a field interested me very much and wanted to kindle the same interest is young children in this region, so after much thought, I wanted to Introduce the children Air traffic and meteorology which will motivate them to take Aviation as their career an develop this field further and prosper in their lives.

I learnt a lot of Information from my seniors in the field of Aviation, where many were experts, Hence any suggestions for Improvement of this effort is well appreciated and it with go a long way in Improving this book. I with open arms welcome all suggestions for further presentation in this work.

I am Radha Krishnan, friend and ex-colleague for a longtime of Shri. Rajusadanand. I met him in Calcutta for first time. And since then we had a nice time together. Rajusadanand since joining in Airports Authority of India as Air traffic controller also served in India Meteorological Department as Senior observer. He has a good knowledge of meteorology. And he has lot of Interest in the field of Aviation and Air traffic control just like me. And whenever

Foreword

we met, we were discussing about Aviation and its development. At civil Aviation training centre Allahabad, we used to spend most of our leisure time in library only. And whenever we used to meet, we used to discuss the latest developments in the field of Aviation.

Chapter 1

HISTORY OF AVIATION

Wright brothers would have never thought that the maiden flight, they conducted in a place called Kitty Hawk of twenty two yards, would be developing into a major field called Aviation Industry with thousands of sophisticated wide bodied Aircrafts like Boeing and Airbus and transporting millions of passengers within the shorter time around the globe, the fastest mode of travel and transportation. No other Industry as developed so fast as Aviation Industry within the span of one hundred years. And this is the first step for man conquering other planets and space.

Right from Ancient times, man has longed to fly in the sky like the birds so that he can view the landscape from the top, in the sky. Even in our Indian mythology, in one of the famous Epic Ramayana. It is mentioned Ravan came in the "Pushpak Viman" to take SITA from the jungles to his kingdom in Lanka. After that in India, we made no efforts to capture the sky.

In the medieval times, the greatest all round genius the world has even produced in History, Leonardo da Vinci who belonged to ITALY, painted the first diagramme of the design of the Aircraft. In the start of the Nineteenth century lot of effort was done for flying.

One Pasteaur Elmer, in effort to fly wore the wings like bird and attempted flying jumped from a building and died. And lot of people tried to fly with balloons and were little successful. Later on men observed the birds body structure and designed that flying machine should be like this. Actually it was wright brothers, who were running a cycle shop, then decided that they will design and fly a powered flying machine. They conducted lot of experiments in the windpipe and designed the ultimate flying machine. When the flying machine was ready, on the beaches of Kitty Hawk in the year 1906, and the first flight was flown. The first flight of twenty-two feet by Orville Wright. And the other brother Wilbur was in the ground and witnessed first powered flight in the world. And the flight of twenty-two feet was first dot on world Aviation map, so Aeroplane was born.

And later on, wright flying brothers improved upon the machine they had invented and demonstrated in front of the public. And people accepted that man can fly from one place to another like a bird. And the sky was at last conquered.

Later on, wright brothers took their flying machine to Europe and they demonstrated their flying machine in big cities like London, Paris and Berlin. And Europeans got interested in flying concepts and immediately the flight by as Aeroplane was smash hit.

Initially people used Aeroplane for small distance flying and for joy ride in the sky and people wanted the thrill of being high in the air and flying and landing. Lot of small accidents occurred in the sky and as usual, man learned from initial mistakes and very important factors to remain the air and fly to long distances. The landing of an aircraft was a very critical phase.

As the first world war about to end, people wanted to engage their new invention Aeroplane in the war. The two world wars played a great role in the development and advancement of Aviation Industry. The aviation field made grand strides to unimaginable level so that people could not believe that this is such wonderful invention and wanted to use in all a fast mode, of transport from one place to another. No doubt the two world wars devastated the mankind, of normal life and civilization but it proved to be a boon in disguise, as because of these two world wars, the faster mode of Air transport with more and more capacity was borne. And that was a boon to mankind in Air transport. Because of the competition to capture Air between the Allied Powers and Axis Powers, developed the field of Aviation.

After the two world wars I & II, people started using the knowledge and experience gained in the wars, in the improvement of Aeroplanes and Air Transport. Aircrafts flying with High speed and Huge capacity, These aircrafts were the result of these two wars.

The Aeroplane were of very high speed and huge capacity to people started using them in AirTransport.

In 1945 the World War II ended with the suicide of Hitler, The Dictator, The United Nations Organisation (UNO) was formed with greater bondage and more membership more than the league of nations which proved a failure and could not prevent World War – II.

International Civil Aviation organization (ICAO) was formed as one of the specialized agencies or organisation of the United Nations (UNO). Once formed, ICAO got into the act of regulating the civil Aircrafts, and their movements all around the world. And many countries where Airtransport was there, became the members of the

ICAO and signatories of the ICAO convention. India also followed the group and became member and signatory of the organization. ICAO wanted civil Airtransport around the world to be regulated and follow some regulations. To enforce this, the rules of air was formed by civil Aviation Experts around the world, and India, adopted all the rules of Air. To enforce this Government of India introduced Directorate General of Civil Aviation (DGCA) an organization to regulate the Civil Aviation in India. DGCA is the regulatory body of Civil Aviation in India. ICAO's headquarters is in Montreal, Canada.

As Civil Aviation was growing one more field of science was also growing. That is meteorology (weather). Initially the study of meteorology was for sailors going to the sea.

In Ancient times the sailors going to the sea were alarmed by the rough seas and turbulent weather. The sailors wanted some knowledge about the turbulent weather development, so that they could avoid rough seas as the ships were tossed about here and there in the rough waters. And the study of meteorology was developing, as Aviation also started. The study of meteorology became necessary for Aviators as well, because like rough weather in seas the rough weather in Air also tossed the aircrafts up and down, so the pilot wanted to avoid turbulent weather, enroute. UNO developed a specialized body called Word Meteorological Organization (WMO) for study of world weather. In India Govt. of India started a organization called India Meteorological Department (IMD) for weather. WMO headquarters is in Geneva, Switzerland.

ICAO wanted the Civil Aviation throughout the world should be uniform and standardized. So that civil Air transport is uniform throughout the world.

Since English is the Language spoken and understood throughout the world, made English has the official language for communication and four more languages used for official communication, they are French, German, Russian and Arabic for enforcing uniform and standardization of Civil Aviation throughout the world. ICAO introduced the Aviation manual DOC-4444, which is the Bible for Civil Aviation throughout the world.

All Civil Aviation practiced, are followed as per instructions from the DOC-4444 (PANS RAC). ICAO has introduced Annexes and SARPS for Civil Aviation designed with Aviation experts all over the world, over there are seventeen SARPS and Annexes and. e.g.

Annex – I– Licensing and Training

Annex – II – Rules of Air

Annex – IV – Aerodrome and ground Aids

Annex – XI – ATS practices

Annex – XII – Search and Rescue.

Chapter 2

NAVIGATION

In ancient times the sailors in the seas, to know the direction they were heading they were using magnetic compass.

Magnetic compass mainly comprised of a magnetic rod suspended freely on a needle. As the property of the suspended magnetic rod, points towards the north-south direction using the magnetic needle, the sailors will find the directions in the middle of the seas and proceed towards the destination they wanted to proceed, using maps. Initially the pilots in the aircrafts were using this magnetic compass to find their way (Navigate) in the vast sky.

In the Indian Air force transport aircrafts along with pilots a navigator also used to be present. His job for was to plot the position of the aircrafts in the chart and guide the pilot with direction they have to fly to reach destination using charts and maps. In a way to Navigate the aircrafts to required destinations.

But nowadays, the commercial aircraft are flown, fly at a very high speed. Their speed is measured in the speed of the sound. So, in the modern aircrafts the post of Navigator has become almost outdated and obsolate. This time people started to use the invention of Italian scientist Marconi's the wireless communication and knowledge of radio waves. This development in the field of communications people thought can be effectively used in Radio Navigation. So they invented NDB and VOR. These equipment have taken

Aviation to another dimension and they utilized in Radio-Navigation, which in Aviation is practiced even today.

Actually Navigation is the process where an Aeroplane or ship is taken from one place to another in the shortest distance and time.

N.D.B. from non direction beacon: It is the one of the Radio Navigation equipment for pilots to guide in the air, there used to be an equipment in the cockpit where pilot can tune a frequency like we do so in radio received.

Every Airport will have atleast one NDB station and the NDB will emit a selected code and the radio equipment in Cockpit, when tuned for particular frequency of the NDB, and will receive the NDB code And when flying over NDB the needle will not be steady and it will be rotating and scalloping that is called cone of confusion, which is certain, that the aircraft is flying overhead the NDB or any other Radio Navigation equipment.

V.O.R very high Frequency OMNI Range: V.O.R is a developed version of N.D.B. From the N.D.B we can find out how near the NDB station with the strength of the radio signal, when the Aircraft is near the station, the signal will be strong and if the station is far off then the signal will be weak. From V.O.R we can find out the radio signal in transmitted from which angle with respect to the aircrafts V.O.R is a very important equipment for the separation of Aircrafts using Radials. Radial separation is used in procedural separation of separating arriving and departing aircraft.

I.L.S Instrument Landing System

It is the radio navigation system which guides the aircraft in the final approach, to the runway with guide slope and centre line of the runway. Pilot has to follow the needle indication of the ILS equipment in the cockpit and it will directly guide him to the centerline of the runway. It reduces the stress for the pilot to visually land in the centre line of the runway and clear of obstacles near the Airport.

I.L.S has got two equipments in the system. One is localizer which guided the aircraft to the central line of the runway. And the other one is Glide Path Indicator (G.P.I) and it will show the aircraft the glide angle (slope) of the aircraft in its final approach to the centre line of the runway. Aircrafts has to simply follow the Glide Path Indicator and Localizer, which will directly guide the aircraft to the centre line of the runway and landing on the runway.

Aircraft landing system using localizer and glide path is called as Instrument Landing System (ILS) with the latest development is Aviation the landing can be done accurately with help of Artificial Satellites (G.P.S) which is called RNAV/ RNP landing system.

Distance Measuring Equipments (D.M.E): It is basically a radar with which the distance of the aircraft from the station (Airport) is found out and mostly, DME is co-located with V.O.R. DME co-located with V.O.R give the distance and radial into which aircraft either approaches the station (Airport), or flies away from the station. Actually a DME co-located with V.O.R is very useful for arrival and departure of aircrafts.

Now in advance countries in Aviation, like US & Australia & Europe G.P.S using satellites is used for finding out Radials and distance of Aircraft, and slowly trying to replace ILS. But in the convention the majority countries in global aviation are using VOR DME for arrivals and departure as it is established the most reliable landing equipment though (land based) for aircrafts.

Chapter 3

RADIO COMMUNICATIONS

As Air traffic increased lot of accidents took place knowingly or unknowingly. The flying Aircraft speed increased to such a great extent, that the pilots had less time to avoid air accidents. To reduce this menace, International Civilization Organizations (ICAO) imposed some restrictions. Very Important in the Aviation rule mainly to impose safety was that all aircrafts while flying in the air was to always to remain in constant touch with around stations or Agencies. Initially the aircrafts flying in the vicinity of an airfield was communicated through colour flags. In nights they used Haldie lamp where communication was done by colour lights mainly red and green.

Later on in aviation people started using Marconi Invention, that is Radio-communication. Radio waves is of two different wavelength one is long wave and another one is short wave. Long wavelength is called H.F. (High Frequency) and short wavelength is (VHF) Very High Frequency. Generally to contact aircrafts beyond two hundred nautical miles from the station, High Frequency (H.F), radio frequency is used. And its range is more than V.H.F (Very High Frequency). V.H.F is very sharp and effective. The wavelength of V.H.F is very short and congested. Air traffic controllers and pilots like this communication as this is very clear and hence no ambiguity. This is also called as line of sight communication.

High Frequency (H.F) is Radio, range is very lengthy. Its range is so far off an Aircrafts flying in Singapore or Kuala Lumpur can contact Chennai Radio (H.F) station. But in this communication, the clarity is not that much as VHF, and intransmission ambiguity is more. Sometimes it is very difficult to understand what the other person wants to convey and many times it is fading.

Nowadays ADS-CPDLC Communication System has come and it is mainly based on Artificial satellites this is called as Automatic Dependence Surveillance and Controller Pilot Data Link Communications (ADS–CPDLC). When Indian government was spending in crores for the space programs in the eighties, people (public) was criticizing it as a wastage, not required for India. Now the orbiting satellites have taken India towards a great step forward, in the field of the communications.

ADS–CPDLC system is Automatic dependent surveillance system and controller pilot data-link. First to operate under ADS–CPDLC systems the aircraft should be capable of logging on to ADS-CPDLC system, in the aircrafts flyings in the cockpit should have ADS–CPDLC system on board.

First aircrafts capable of ADS-CPDLC link system, the logs in to ADS-CPDLC system in ground. So they initiate in the ground system and when the periodicity rate is corrected. The aircrafts get connected in ADS and the aircraft and ground system has to set contracts and the first one is Periodic contract and the default rate is 1680 seconds, second one is Event contract and the third is Demand contract. And the CPDLC is Controller Pilot Data Link and the controller pilot message are sent through the computers of ground system just like messages in the SMS.

(G.P.S) Global Positioning System

This is called as Global Positioning system and this is the constellation of artificial satellites sent by America to space and circling around the earth in orbit.

Glonass: It is the constellation of artificial satellites sent by U.S.S.R (Russia) orbiting around the globe. G.P.S. and GLONASS are constellation of artificial satellites orbiting around the globe actually twelve satellites selected in GPS and twelve satellites selected in GLONASS form the ADS–CPDLE system. Because of the system any aircraft flying not in the Radar coverage areas, can be seen and monitored in ADS-CPDLC screen, ADS-CPDLC is Automatic dependence surveillance system and controller Pilot Data Link with ADS system civil can find Aircrafts location and altitude which it is flying around ground system (ADS-CPDLC) sends messages to cockpit which is responded by the SMS (flight manager system) in the cockpit.

(F.M.S) Flight Management System

Initially in the cockpit operated by the pilot the panel was containing Altimeter to find out the height the aircraft was flying and pitot tube to find out the speed of the aircraft which was called ASI is Air Speed Indicator and then fuel meter, as displayed in the motorcar, so that pilot can find out the fuel position, Artificial Horizon and also the control of the aircraft. Nowadays the cockpit has so many Instruments which are mostly electronic and an are displayed in a computer called F.M.S Flight Management

System, the pilot operate F.M.S like a computer. And F.M.S co-ordinates with the ground control system as ADS-CPDLC and send all informations to the ground system without even the knowledge of the pilot and also in the cockpit is the Auto pilot system which takes control after an aircrafts attains the cruising level, controls the Aircrafts.

ICAO has standardized the radio communications (mainly transmissions) between the flying pilots and the ground stations, which is called ICAO phraseology controller and the pilots should comply with this format 99% in their radio communications.

Chapter 4

ATS-SEPARATIONS

As Air traffic increased the number of accidents also increased, endangering the safety of the flyers. ICAO started working out ways and means to reduce the number of accidents to the minimum or null level. ICAO designed procedures and separations for the civil aircrafts to follow with the panel of experts in aviation who have taken account of the behavior of the different types of aircrafts and weather occurring in the sky.

ICAO entrusted this responsibility of enforcing rules and regulations is civil Aviation to ATC (Air Traffic Controller). Air Traffic Controllers enforce the separations laid down by ICAO for the safety of aircrafts and also frequently monitor the operation with the equipments they are provided with VHF, HF, Radio, Navigations aids, ADS-CPDLC system and Radar.

(Contents taken from ICAO-Doc 4444)

(V.S.M) (Vertical Separation Minimum)

It is supposed to be the best separation in civil Aviation designed by Civil Aviation Experts. Vertical Seperations is obtained by requiring the aircraft using the prescribed altimeter setting procedures to operate at different levels expressed in terms of flight levels or altitudes in accordance with the altimeter setting procedures.

(VSM) Vertical Seperation Minimum

Vertical Seperation Minimum shall be a nominal 1000 ft, below FL-290, above FL-290 upto FL-410 is nominal 2000 ft. Vertical Seperaton Minimum shall be a nominal 1000 ft between FL 290 and FL 410 both inclusive in RVSM airspace.

ATC Unit shall normally authorize only one level for as aircraft beyond its control area.

The eastbound aircraft will maintain odd level whereas west bound aircrafts will maintain even level i.e., Aircrafts from India to Bangkok, Singapore, & Kuala Lumpur with maintain F 310, F330, F350 the aircrafts coming to India from Singapore, Bangkok and Kuala Lumpur will maintain even levels like F-320, R-340 & F-360 etc.

Lateral Seperation

By reference to the same or different locations by position reports which positively indicate the aircraft are over different geographic locations as determines visually or by reference to a navigation aid, by use of some navigation aid or method.

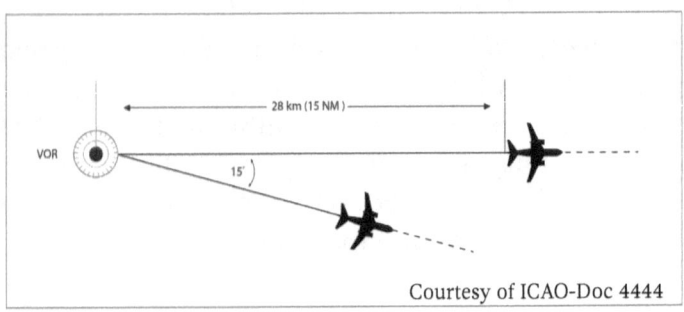

Courtesy of ICAO-Doc 4444

VOR: both aircraft are established on radials diverging by at least 15 degrees and atleast one aircraft is at a distance of 15 Nm on more from the facility.

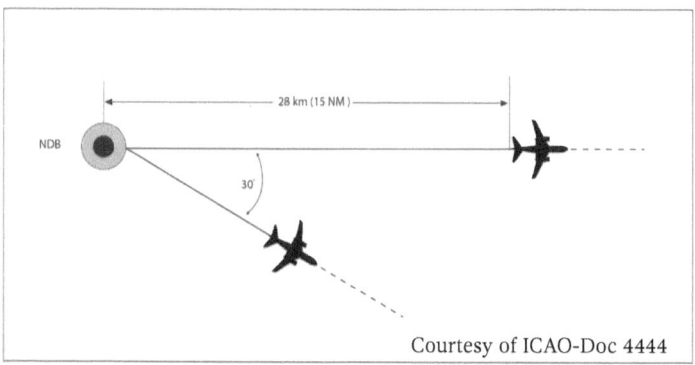

Courtesy of ICAO-Doc 4444

NDB: both aircraft are established on tracks to or from the NDA which are diverging by at least 30 degrees and at least one aircrafts at a distance of 15 Nm or more from the facility.

Dead Reckoning (DR): both aircraft are established on tracks diverging at least 45 degrees and at least one aircrafts is at a distance of 15 Nm or more from the point of inter section of the tracks this point being determined either visually or by reference to a Navigation aid and both aircraft are established our bound from the inter sections.

Longitudinal Separation

Longitudinal Separation based on time.

Aircraft flying on same track

- 15 minutes or
- 10 minutes of navigation aids permit frequent determination of position and speed

Control Tower & Weather

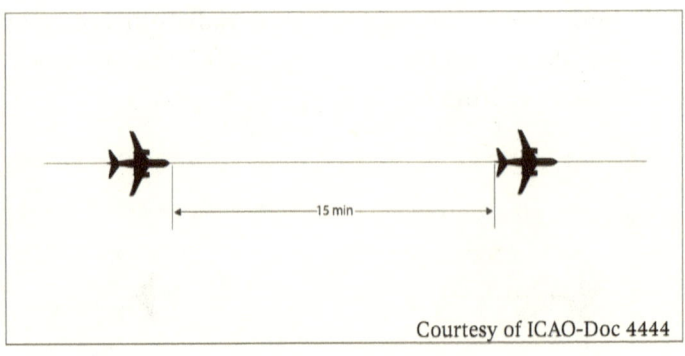
Courtesy of ICAO-Doc 4444

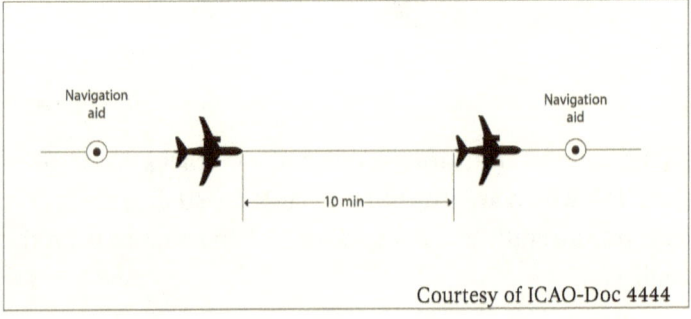
Courtesy of ICAO-Doc 4444

Aircraft flying on crossing tracks:

- 15 minutes at the time of inter section of tracks
- 10 minutes, if Navigation aids permit frequent determination of position and speed

FL – Flight Level.

Aircrafts climbing or descending

Aircraft on the same track when as aircraft will pass through the level on another aircraft on the same track the following minimum longitudinal separation shall be provided

- 15 minutes while vertical separation does not exist

- 10 minutes while vertical separation does not exist provided that separation is authorized only where Navigational aid permit frequent determination of position and speed.

Aircraft on crossing tracks:

- 15 minutes while vertical separation does not or
- 10 minutes while vertical separation does not exist of Navigational aids permit frequent determination of position and speed.

Aircraft on reciprocal tracks:

Where lateral separation is not provided, vertical vibration shall be provided for at least ten minutes prior to or after the time the aircraft estimated to pass, or are estimated to have passed.

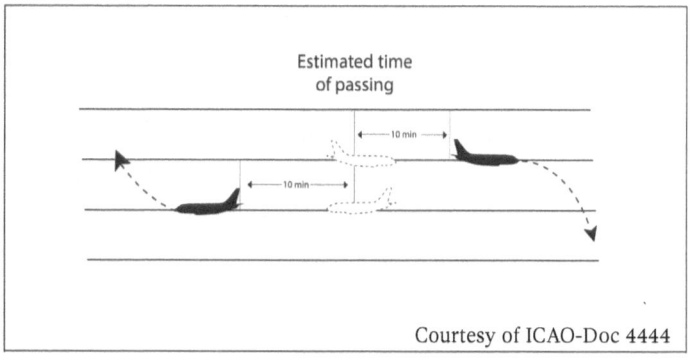

Courtesy of ICAO-Doc 4444

Aircraft climbing or descending on the same track:

10 NM while vertical separation does not exist provided

- Each aircraft utilize on track "DME stations.
- One aircraft maintains a level while vertical separation does not exist

- Separation is established by obtaining simultaneous DME readings from the aircraft

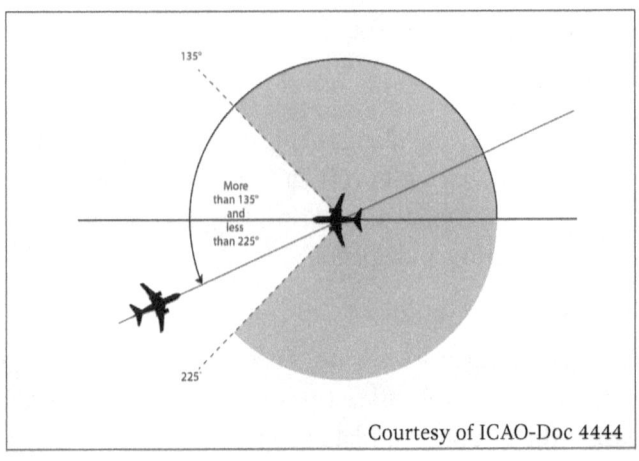

Courtesy of ICAO-Doc 4444

Aircraft on reciprocal track:

Aircraft utilizing on track DME may be cleared to climb or descend to or through the levels occupied by another aircraft utilizing on track DME provided that it has been positively established that aircraft have passed each other and are at least 10 NM apart or such other value as prescribed by the appropriate ATS Authority.

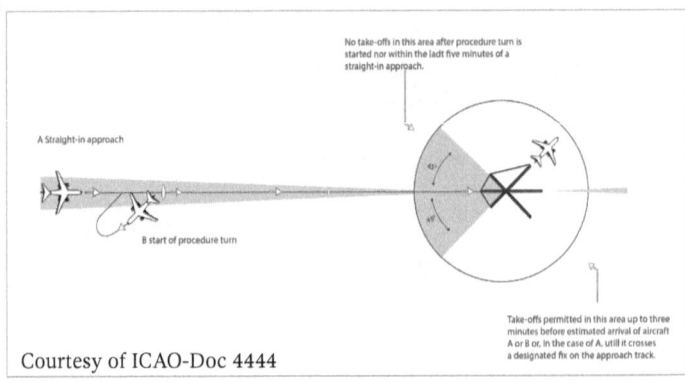

Courtesy of ICAO-Doc 4444

If an arriving aircraft is making a complete instrument approach, a departing aircraft may take off.

- In any direction until an arriving aircraft has & started its procedure tun or base turn leading to the final approach.
- In a direction which is different by at least 45 degrees from the reciprocal of the direction of approach after the arriving aircraft has started a procedure turn or vase turn leading to the final approach, provided the take off will be made at least three minutes before the arriving aircraft will be in the beginning of the Runway.

If an arriving aircraft is making a straight is approach, a departing aircraft may take off

- In any direction until five minutes before the arriving aircraft is estimated to be over instrument runway.
- In a direction which is different by at least 45 degrees from the reciprocal of the direction of approach of the arriving aircrafts
- Until three minutes before all the arriving aircraft is estimated to be over the beginning of the instrument runway.

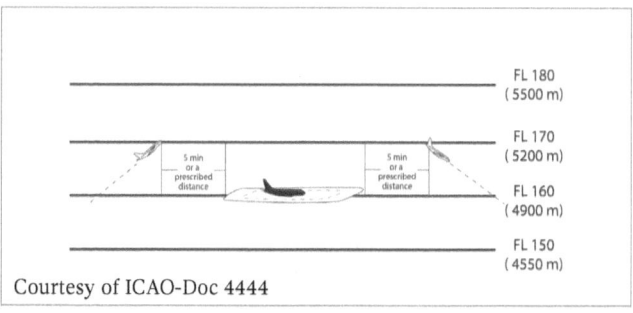

Courtesy of ICAO-Doc 4444

In Aerodrome Control Tower Jurisdiction

A landing aircrafts will not normally be permitted to cross the runway threshold on its final approach until the preceding departing aircraft has crossed the end of the runway in use or has started a turn or until the preceding landing aircraft are clear of the runway in use.

A departure aircraft will not normally be permitted to commence for take off until its preceding departing aircraft has crossed the end of the runway in use, or has started a turn or until and preceding landing aircraft are clear of the runway in use.

Chapter 5

TRAINING

The back bone of Aviation is the training. And this is not an exaggerated statement. And the back bone of any organisation is, the training unit. It is said that the battle of Wellington was won on the playgrounds of Eton. Training gives the chance to correct the errors and which cannot be committed in the live field (esp Aviation) the core motto of any Aviated related agencies is the safety of passengers and crew of an Aircraft.

So there is no compromise in the standard of training in the field of Aviation either it is pilot training or it is the Air traffic controllers training. Generally in Air traffic controllers training which is of many parts (many stages). The controller trainees are introduced to Aviation and meteorology technical equipment and effective communication methods. The longest is the practical training where the trainee is given training to cope up with Aviation speed & emergency situations.

The Air traffic control training mainly focuses on improving the ability of the controller to cope up with the density of Air traffic without compromising the safety of the aircraft. It is the kind of real situation in the control tower simulated (created by the trained instructors) which improves the reaction of the controller to cope with the normal traffic and to tackle the situation where unusual (abnormal) occurrences in the aircraft which the trainee is taught to tackle without hampering the safety of the aircraft. And this gives the controller, confidence to tackle the situation in real life in Airport. The training is so tough the controller gains confidence. A controller is trained to react correctly to realise situation in air control traffic limits.

Initially there was one training centre in India at Baumrauli near Allahabad, which is in Uttar Pradesh (UP) but now due increase in the number of training programmes two more training centres developed at Hyderabad and Gondia in Maharashtra. The pilots training for Air India is in Begumpet near Hyderabad.

Chapter 6
MODERNIZATION OF AIR TRAFFIC CONTROL ESTABLISHMENT

In the early Nineteen Eighties (1980), computer was introduced in India. Till that time in India everything was conventional for example in education, it was conventional method, using blackboard and same teacher teaching most of the Subjects will be teaching English, Science, Mathematics, History and geography and the local mother tongue of the region. In all administrative officer and banks, it was files and Registers and Account books.

With the advent of computers, all the datas were stored in the computer Airlines and Railway tickets booking and reservation changed an Air traffic control services also the entire system was changed mainly communications previously all the flight details were transmitted through teleprinters and telex and through telephones and STDs.

Now all the Datas mainly the flight details of different aircrafts operating were stored in a system and transmitted to the Airports throughout by Automatic Switching System (AMSS).

And previously the flight progress strips were prepared manually. And a team was working on the preparation of flight progressions strips with the help of the flight plan and the strips were updated, as the flight progressed. Now all these prepared into the system with the help of Automation systems. Now in major Airports in India the Raytheon system is used which integrates the radar display in all major Airports in India which provides continuous radar coverage in the land area. And in sea and ocean areas the ADS-CPDLC system is used for coverage of flight progression display, surveillance and communication. And the GPS – GLONASS Satellites integrated with Indian satellites which augments the display and the accuracy of the position of the aircraft is increased.

After the advent of the ADS-CPDLC in the entire ocean stretch the Air traffic display could be observed and this was very useful in sorting out the traffic in Oceanic Region. The number of Incidences which happened in manual display board scenario reduced drastically as the Aircraft movement display was observed. The main principle of this system is the GPS system using satellites controllers could communicate with the aircrafts equipped with ADS-CPDLE system in the cockpit FMS could communicate their requirement with the controllers and get the facility and suitable level which decided their fuel economy. Now of late the latest ADS-B system where the aircrafts can be descended and climbed and traffic could be sorted out like in Radar sceneries their ADS-A system replaces the Radar system in the Oceanic Scenario in surveillance of the traffic. And the computers and the printers totally replace the manual system in FPS. An Air traffic controller should be always alert and workout any problem in the airport.

Chapter 7

ROLE OF FIRE SERVICES IN AVIATION

Fire services play a crucial rule in the safety of the Aeroplanes and the Airport. ICAO has insisted that every Airport should have a devoted fire services for the safety of the Aeroplanes and Airport equipments. Since the Aviation fuel is pure kerosene and highly inflammable. Any Aeroplane crashing or has heavy impact on the ground immediately catches fire and the fire quickly spreads to the Airport installations and the personnel working in that. The safety of the lives of airport personals also to be taken care of.

So every Airport has established a very strong powerful dynamic fire services vehicles consisting of very well trained fire personnel. who are ever ready to proceed to the affected part of the Airport at the earliest and put off the fire within the shortest possible time. They are trained like defense, military and police personnels. They are all very good drivers and take their crash fire tenders carrying fire existinguisher and control the fire spreading our and extinguishing them.

To extinguish fire the fire personnel should know the properties of fire and the nature of fire. The fire is a triangle that is formed of three contents. It consists of fuel heat and oxygen. These three factors contribute fire.

The material for catching fire the oxygen present in the atmosphere and third and last one is the heat. And if we remove one of the contents, the fire is extinguished (put off). The Fire personnel act this principle when controlling the fire from spreading and putting off in the shortest possible time their minimizing the loss of materials and human lives. And widely using water the cooling action is done. And by the probes of smothering (In which the soap water is used in doing a carpet act and the oxygen in the atmosphere is cut off.

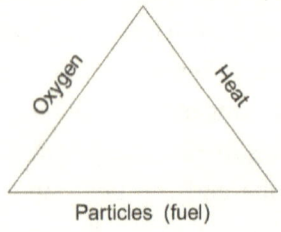

Particles (fuel)

Every Airport fire station is rated by the category. The category of the fire station depends on the widest wing span and the largest type of Aircraft and the biggest Aircraft operating in the Airport for e.g., A380 jumbo then the category of fire services has to cater to the B747 (JUMBO) in the Airport.

The fire personnel are trained by training centres in the Calcutta (Narayanpur) and Delhi Palam Airport the fire and Rescue operator is well trained to extinguish (put off) any type of fire (solid, liquid and gaseous the fire personnel are fit like Army and police personnel. And are expert drivers of heavy vehicles upto ten gears (five crash tenders). And they are trained be reach the affected spot in the Airport in the shortest possible time called the critical time. And reaching the spot they are supposed to get into act and control put off the fire services not only cater to the fire in the Airport vicinity in also supposed to provide the support services to the nearby fire mishaps happening anywhere in the nearby towns or villages near the reachable limits say within 50 kms around the Airport. Since the Airport fire service personnel well trained in fire controlling activity they perform well outside also and lead the rescue services.

So the fire station in the Airport not only as asset to the Airport but also to the public in the society as fire mishaps any user in the society causes lot of financial loss but also loss of human and animal life even that's why the Government through the Airport Authority management has purchased sophisticated transport (that is costly fire crash tender (CFJ) and sophisticated latest firefighting equipments. So that any fire mistake is controlled at the earliest.

Aviation fuel catches fire in an aircraft immediately on any slight impact or collision with any obstacle either ground or any installations.

Chapter 8
METEOROLOGY

For pilots the knowledge of meteorology is as important as the knowledge of road is to a driver. As motor vehicles surface transport depends on the road conditions it travels and ship transport depends on the sea water conditions. Air transport depends on the sky and weather conditions and this is why the meteorology knowledge is very vital for the pilots.

In sky of the 80% gas is Nitrogen and 18% is composed of oxygen gas. And insert gases Neon Xenon fluorine, chlorine and Krypton and water vapour consists of other 2% of the Atmosphere.

Our Atmosphere is divided into four main parts, they are, Troposphere, Stratosphere, mesosphere and Ionosphere. We live in troposphere, and all the main activities of the atmosphere like rain, shower, lighting, thunder, storm and cyclone all happened in the Troposphere only.

Troposphere in the zone from the ground to twelve (12) kilometers high.

Troposphere in the equator stretches up to twelve kilometers (12) height but while moving towards the polar region it shortens to (8) kilometers. The top most layer of troposphere is called tropopause most of the weather process that happens in the earth takes place below the tropopause layer. The temperatures near the ground is always very high and when we go up the temperature reduces and near the troposphere, at the top mark layer of the troposphere is almost in minus degrees. The Jetstream (band of wind) occurs just below tropopause level.

Stratosphere

The next zone in the atmosphere is Stratosphere. The temperature which drastically dropping in troposphere as per attitude suddenly starts increasing as per altitude. In this layer, the U.V. rays of the sun are filtered out and U.V. rays filtered by the Ozone (O2) layer. The U.V. rays are very bad for our skin and the Ozone layer in the stratosphere thus protects in the outer skin. Because of this ozone U.V. filtering action, in this layer the temperature rises.

Mesosphere

The next layer in the atmosphere is the mesosphere. In the mesosphere the temperature which was shooting along with the altitude drops gradually. The graph goes on dropping and it becomes too much cooling.

Ionosphere

The last layer in the atmosphere is the Ionosphere. In Ionosphere the temperature which is dropping in the mesosphere rises again. In the Ionosphere, the atmosphere is split into ions. In this layer which is 'E' layer, the radio waves gets deflected and travels miles and miles ahead. This is the reason, the radio waves emitted by the aircraft in Singapore is picked up a radio station in Chennai. The Ion layers acts as a sheet when sunlight and heats is more and becomes thick. This time radio frequency can be low, which can be picked by the receiving radio station. But in night time when the Ion layer is less, due to no sunlight and heat is low, than this time the capacity of the radio emitting signal is increased in frequency so that the distance radio station receives it signal that's why in night over the ocean and sea the radio signal reflected is strong. The radio signal bounces because of reflecting in 'E' layer present in the Ionosphere.

In atmosphere where we live the Air very important for our livelihood. Atmosphere contains 18% of oxygen, which is very important for life, we breathe the oxygen in the atmosphere which is very important for our metabolism. The oxygen, we breath in purifies our blood and burns out energy which is very important for our normal daily metabolic activities we living beings, inhale oxygen and emit the carbon di oxide which is useful for plants for preparation of its food.

Air appears to have no weight, but it has weight. To verify this, we have to have a empty balloon and weight it. And again fill it up with air and weigh it again. The weigh difference will gives the weight of the air. Thus we can understand that the air has weight.

Air mass behaves much like liquid, As, when we fill up a glass jar with water (liquid) When we put a hole in the top of the jar and also a hole in the bottom of the jar we would find the water flowing but in the bottom is more. In the same way is the atmosphere, air is also very dense in the bottom. That is near the ground the air density is more and oxygen is also more. As we go higher up, the air density decreases this property or air in the atmosphere, is used for calculation of altitude at which an aircraft flies in the air. The find the density and pressure of air an instrument called Barometer is used.

Chapter 9
PRESSURE

Barometer

The Barometer is an instrument used for measuring air pressure the altimeter is used for measuring the height at which an aircraft flies. And an altimeter is basically a barometer calibrated for height. In the starting mercury barometer were used. That is basically mercury filled in a calibrated glass jar which is tilted upside down in a container. The glass jar is calibrated so that the atmosphere pressure is measured. Nowadays the mercury barometer is replaced by aneroid barometers.

Altimeter (Aneroid barometer) helps in calculation of the flight level at which an aircraft flies and it is very useful in maintaining the vertical separation minima and it is useful for Air traffic controllers in the separation of aircraft and allowing the aircrafts to fly in high level which minimizes the exhaustion of Aviation fuel. Thus it saves the economy of the Airline operators. As the aircrafts flying in the higher level consumes less fuel.

The Regular Atmosphere pressures of different stations are collected from the meteorological observations periodically every three hours and is utilized by the meteorological analyst in drawing Isobaric lines of (same pressure) with which in map they get information where the low pressure area exists so that the weather phenomenon occurs and warn the public and the fishing communities about the location of cyclone.

Control Tower & Weather

Whenever in an area where low pressure exists air mass from surrounding areas rush in to make its normal just like water mass run from high area to low areas. And strong cylones are created over seas and winds carry lots of moisture (water mass).

Generally, in literature the wind is supposed to be used as an adjective to indicate the strengths of a person. It is always associated with sports where using physical strength is the crucial factor. In Indian mythology, the puranas the strong fighters of both Ramayana and Mahabharata are supposed to be of the sons of God of wind. Actually, the phrase he swings like a wind in the combat indicating the speed of a person. Bhim and Hanuman supposed to be sons of Vayu. There was a story that long time back a warrior fought like a tiger in battle field. And in the end of the day he was wounded to the hill and was on death bed, the goddess of victory was impressed and pleased by the bravery, appeared before him and asked him he wanted before his death. His only request was breeze. After he died, gentle breeze flew over his body. This story indicates the pleasantness of gentle breeze, there is an attempt to all air also, which we all should be people who welcome a trough i.e., low pressure depression which moves very slowly and there is plenty of rain and land area is benefitted, Whereas in cyclone the wind is a strong factor and devastation is heavy as houses are blown, trees are uprooted and electric lamp posts are bent and worn our wires cause electrocution of living beings and cultivated lands are destroyed.

Actually during day time the sun heat both land and water (sea), and the land gets heated faster and the water gets heated slower. So the land is hot, so the wind blows from sea to land (i.e. Sea Breeze). And in the night time the land cools faster than the sea (water). So, in the night time the wind blows from the land to sea (i.e. land Breeze).

In the mountains, there is windward side and leeward side. In the windward side, there is plenty of wind and in the leeward side the wind flow is less.

Chapter 10

TEMPERATURE

Actually, the sun's heat is the main cause of the weather, all over the world. You will be surprised to hear that. So, because of sun's light and earth's rotation on its own and the day and night happens. Because of sun rays, our skin gets the vitamin 'D' energy and our skin becomes thicker and stronger. In the ancient days people worshipped sun as god. In a way, that is correct, just imagine a world without sun, then we will understand the importance of sun in the solar system. Even for our everyday life sun and sunlight is very much essential. And because of sun only the plants cook their own food and inhale carbon dioxide which we and other living beings exhale and emit oxygen, which is very essential for survival of living things. Temperature is the creation of sun's light and sun's heat. At the day progresses sun heats up land, water and everything in the earth.

The sun rays first heats the atmosphere and then the earth landmass and water. So, since the earth is heated, the temperature highest near the ground and as we go up the temperature reduces and this is called as lapse rate. And we expect the temperature to the highest at mid noon (12'o clock). When the sun is high on top. Actually, this is not so, the sun first heats air and then earth. And the heated earth starts radiating heats, which is called terrestrial radiation. The sun rays and the heated earth combined to create the highest temperature on the day at four 'O' clock (i.e. 4'o clock). The same way the heated earth starts cooling in the night and the atmosphere also starts cooling and as usual, we expect the lowest temperature and the coolest at around midnight twelve 'o' clock but this is not so. The lowest temperature of the day is recorded just before the sunrise (i.e. around 5'o clock) when earth and atmosphere have cooled to the minimum temperature. The temperature as we go up in the troposphere drops and this is called as the lapse rate. And the temperature drops till we reach the topmost layer in the troposphere.

Normally in a sunny day, the sun heats up the land and water sources and the heated up water becomes vapour and forms a separate entity and the water vapour mass becomes a separate entity and as it goes up, loses temperature and becomes a cold water mass which is called clouds and supercooled water and the heat lost by the water vapour is called as adiabatic lapse rate.

The water vapour collected together or cling together and it is like water vapour mass and behaves like a balloon filled up with air. And it cools down as the vapour goes up in the atmosphere and it loses heat which is called adiabatic cooling. When the supercooled water in the cloud becomes heavy so that the cloud cannot bear then it diffuses and disintegrates and because of this precipitation of water (rain) takes place.

Chapter 11

TYPES OF CLOUDS

About clouds, in Indian mythology there are many stories one of them is gods used to roam over the clouds and Lord Indra is supposed to be the god of clouds. And only when he wishes there will be rain other ways the earth will become dry. People pray to him to bring rains to drive away drought. Otherwise severe water problems like, we faced in Chennai this year (2019).

But scientifically, due to sun heating the water mass in the land and sea, water gets heated up and becomes water vapour. And mass of water vapour cling together to form the clouds. When more water mass accumulates, the clouds becomes dense and heavy.

Clouds as per their height are categorized into three parts they are high clouds, medium clouds and low clouds etc.

Low Clouds

Stratus cloud

It starts from five hundred (500 ft.) to three thousand feet (3000 ft.) In earth, wherever there is a climate change and rain occurs, it should be due to low clouds. Stratus clouds is the lowest cloud and generally the height is 300 ft up to 1000 ft. It appears like the cotton sponge and moves very fast. It is generally observed during the bad weather times and especially during the formation of low pressure area, trough or the cyclone time there is too much rain in the places. The low clouds can be observed moving very fast.

Stratocumulus cloud

The next layer of low cloud is stratocumulus cloud and it starts from height 1500 ft up to 2500 ft stratocumulus (SC) cloud darkish in colour and more definite shape. Stratocumulus generally appears in the morning and evening and during rainy days when stratocumulus is moves in the sky then it is a indication of a rainy day and bad weather stratocumulus moves slower than stratus. Stratocumulus clouds is more in bad weather. Stratocumulus clouds also brings plenty of rain.

Cumulus

Cumulus clouds are shiny white collyflower shaped clouds which appear in the sky in the morning and afternoon and in fair weather when the sky is blue. It will be a surprise to note that the dreaded cloud (i.e., cumulonimbus (CB) cloud is an offshoot of cumulus. Cumulus clouds with vertical buildings on a rainy day, later becomes cumulonimbus. Generally in the morning we can find there is fast vertical build up in the Stratocumulus clouds.

The Medium clouds

The Medium clouds consists of Altostratus, Nimbostratus and Altocumulus.

Altocumulus

The most popular among the medium cloud. Its height is from 8000 ft to 12000 ft. Its shape is like and bubbles in the sky. The altocumulus colour is white and its appearance like sand bubbles and it extends in large part of the sky. Sometimes there is rain from Altocumulus clouds also. Its appearance in the sky is like muddy sand accumulation.

Altostratus

It is a thick white sheet of cloud completely covering the sky during the monsoon rainy seasons. The presence of Altostratus in the sky is a sure indication of the bad weather. Generally when Altostratus is observed, sky is overcast and bad weather and rainfall is there.

Nimbostratus

This clouds are dark clouds occurs only during very bad weather, i.e., whenever cyclone occurs and heavy downpour of rain occurs. This cloud contains plenty of moist present in them. They are mainly collection of cumulonimbus clouds with lot of lightening and thunderstorm activities acting in a cluster. When nimbostratus cloud exists the precipitation (rain) may even continue for two or three days.

High clouds

Cirrus: - These are high clouds and generally exist as around thirty thousand feet (30000) to thirty-five thousand feet (35000). Cirrus are thin layer of pony tail shaped clouds in the clear blue sky. They leave trail of white clouds which are thread shaped and appear very beautiful in the blue sky. Generally they appear in the fair weather or clear sky. It appears like small droplets of water in the flow.

Cirrocumulus - It is altocumulus clouds at thirty-five thousand feet (35000 ft). It is not as distinct as altocumulus cloud and it appears as muddy water sprinkled of water. Its appearance in the sky is just like that.

Cirrostratus: - It is same as altostratus cloud formed at an height of thirty-five thousand feet (35000 ft) or at forty thousand feet (40000 ft). It is the highest level any cloud exists. It appears exactly like a thin sheet of glass in the sky. Sometimes due to cirrostratus precipitation (rain) occurs. And it is generally like slow or less.

Cumulonimbus clouds: - Now we can see the most important low clouds, the cumulonimbus clouds. Popularly shortly called as the 'CB'. It is the most dangerous of all clouds.

The cumulonimbus ('CB') clouds originates from the cumulus (cu) clouds.

In summer season in the morning when it is fair weather the cumulus form and appear in the sky as cauliflower buds. In the mid afternoon when the sun reaches the top of the sky the heat radiated is maximum and the cumulus clouds build up a tall towering structure and shining white colour and there is lot of activity inside the clouds. And in the evening the towering cumulus structure becomes anvil shaped and starts elongating sideways and there is lot of positive, negative charges formed in the water bearing clouds. There is updraft and downdraft of air inside the cloud. Because of that, lot of turbulence inside the cloud.

When Cumulonimbus cloud is fully formed it is of anvil shaped, and when it is broken it dissipates into a cold air and blows in the opposite direction and there is a cold air wind and super cooled water inside the cloud precipitates as rain and it results in heavy wind and heavy rain. And lot of lightning and thunder storm activity occurs.

The aircraft, whenever gets the information of the position of Cumulonimbus (CB) cloud, should avoid entering, as when it enters into a strong 'CB' cell, it is to used up and down due to updrafts and downdraft of air column moving inside which will toss the aircraft up and down and make its fly path very rough and tough. Pilot flying is these situations will find it difficult to bring the plane under control. And the passengers flying inside will experience uncomfortable flying.

The electric charges negative (-ve) or positive (+ve) will spoil the aerodynamic structure of the aircraft. The super cooled water in the 'CB' clouds forms Icing on the aircraft (mainly wing panels) and increases the drag of the aircraft. The Icing also causes the structural damage of the Aircrafts. The aircraft becomes heavy structurally and flying becomes difficult. The cluster (collection of CB clouds) is called Line Squall and it continues for kilometers together and whenever an aircraft enters a Line Squall it has a very tough time flying and is tossed continuously. And generally aircrafts plan in advance, with the help of route weather forecast, and avoid entering into the Line Squall in the route.

Chapter 12

WIND

In the atmosphere we have already seen that there is 80% of Nitrogen and 19% of oxygen and 1% of inert gases we already know earth is a part of solar system with sun as centre an solar system the member (planets) rotate themselves one rotation and revolve around the sun. Earth's rotation on its own is called day and the tirol is takes to revolve around the sun is one year.

For all the weather phenomena happening, the sun and its solar heat is wholly responsible. The Earth's atmosphere is heated by sunrays is a different way which causes the pressure difference in the atmosphere and which causes formation of High pressure and low pressure so there is a pressure variation in the Atmosphere. As water has got a nature of moving from High to low place. Atmosphere also behaves likes a liquid and flows from High Pressure to low pressure to maintain equilibrium. And this causes the wind flow from one place to another. The wind starts welcoming.

When we take earth sphere the Atmosphere height is more is 12 kms and is poles it is less in 8 kms). So in Earth sphere the pressure is high (H) near the equator. And the wind flow is from equator towards the poles i.e., north of south polar regions. Now we should keep in mind the earth is rotating around itself in its own axis is from west to east in a repeating way. So, the wind direction also gets drifted instead of north, its moves towards north-east in the northern hemisphere and exactly opposite in the southern hemisphere. And the effect of Earth's rotation on its own on the winds flowing from equator to polar region is called the coriolis force. And Earth different terrestrial surface also has effect on the wind direction.

And the magnet in the earth from north pole & south pole also has effect on the direction and force on the wind. And the earth's crust with terrestrial variations also has effect on the wind. So, the wind acted by these forces is called as Geostrophic wind.

So, the conclusion is that the wind direction of the wind we experience is not the exact direction of the wind when it started and there are so many forces that act on the wind that changes the direction of the wind. And the temperature variation has its effect in the direction.

Buy Ballot's Law: - Buy ballot one of the world famous meteorologist says that when you face the wind force with your back towards the wind, the left hand side less the low pressure and the right hand side the high pressure. And the wind normally blows from the high pressure to low pressure.

The wind contains oxygen which is very important for survival of human beings. Even in hot summer, when you experience very good force of wind, you don't feel the heat. The heat becomes bearable when there is good amount of wind. The wind becomes a storm, then it causes heavy damage to buildings vehicles and the electrical lamp posts and to the trees or garden. And our Government administration takes lot of action to reduce the damage.

Always the pleasant wind is welcomed, when wind contains good amount of water it is peasant and which it is dry it even heats the body.

The Jet streams (a band of wind) is observed in the atmosphere, just below the tropopause in the troposphere and Clear Air Turbulence (CAT) is observed in the sky without clouds.

Chapter 13

INDIAN CLIMATOLOGY

Weather is a temporary phenomena at a place. But whereas climate is the pattern of the weather expected to be existing at a place at a particular period of the year taken into consideration of observation by the meteorologists for many numbers of years in their maintained records. The climate of a region is the reference for anticipating the weather pattern and the directions of the normal wind supposed to exist, weather observation utilized is noted and wild for future climatological reference.

The Indian climatological study is done by experts years together and arrived at a conclusion, which provides help to the Government of India to prepare its plans for improving the agriculture and to caution and steps to reduce the casualty of human lives due unexpected adverse weather conditions, and Natural calamities like cyclone, tornados, rain and floods, and hot the drought condition like do extreme summer weather conditions, and assessment of the situation expected in the country by the government for its planning.

Indian climate is divided into four seasons

- Winter
- Pre-monsoon
- Monsoon and
- Post-monsoon seasons.

Control Tower & Weather

Winter

In India winter starts from November and extends upto March. This is the coldest period experienced by major countries in the Northern Hemisphere.

In winter the low pressure exists in the southern peninsula that is, in Bay of Bengal near South India and Ceylon. There is a permanent low pressure in the Bay of Bengal sea. Because of that lot of weather and rainfall is experienced in South India.

In North India, the gangetic plains reel under the extreme cold conditions. The maximum temperature to 20-25°C and in the night the lowest temperature reaches up to 5°C. In the Himalayan region, it reaches below minus centigrade. People get inside home in the night and like to move in the day only. The visibility also goes down and the buses and train get affected by long delays, because of the fog. Even in Aviation the flights get delay due to poor visibility and foggy conditions. The international airport Delhi is most affected by the dense fog. Most of the flight schedule is rescheduled due to heavy delay in aircraft arrival and departures.

In the month of January, February a low preserve system formed north of Afghanistan moves towards Pakistan and Punjab plain which brings winter rain to Indo-Gangetic plains and in the Himalayan regions, there is snow fall inmost places. Shimla, Kulu-Manali and in Jammu and Kashmir, during these times people in Delhi and other north Indian cities rush to these places to enjoy snow fall and play in snow with snow balls and enjoy other winter games. The entire Himalyan range is covered with snow.

Generally, in Northern India, winter is the time many auspicious activities like betrothal and marriages takes place. And people consume hot food and beverages to keep them warm.

Pre-monsoon Season

Pre-monsoon season in India starts from the month of March to June. It is the summer season as in the advent of the month of March the sun moves from the Tropic of Capricorn $23\frac{1}{2}$ °S to the Tropic of Cancer $23\frac{1}{2}$ °N. The sun starts moving north of equator after the Equinox (march 21st). And the summer starts in the Northern hemisphere.

The Climate all over India is hot and torrid the sun appears early in the morning and it is fair weather with blue sky in the background and buds of cumulus clouds. In the late afternoon cumulus with lot of buildup of clouds becomes towering cumulus and later on in the evening Anvil shaped cumulonimbus clouds is formed on the evening it is followed by strong winds from the cumulonimbus cloud that is attcking land bringing down heavy downpour rain showers (Heavy Precipitation) and sometimes Hailstorms also within the short span, the rain shower fall is heavy. Pre-monsoon weather effect is felt all over India. In North India, the thunder clouds and formed in the Thar desert and it creates sand dunes, which continues as 'Loo' (i.e., dust storm) in Indo-Gangetic plain and spreads all over North India in the summer afternoons. And the dust storm spread hovers in kilometers height.

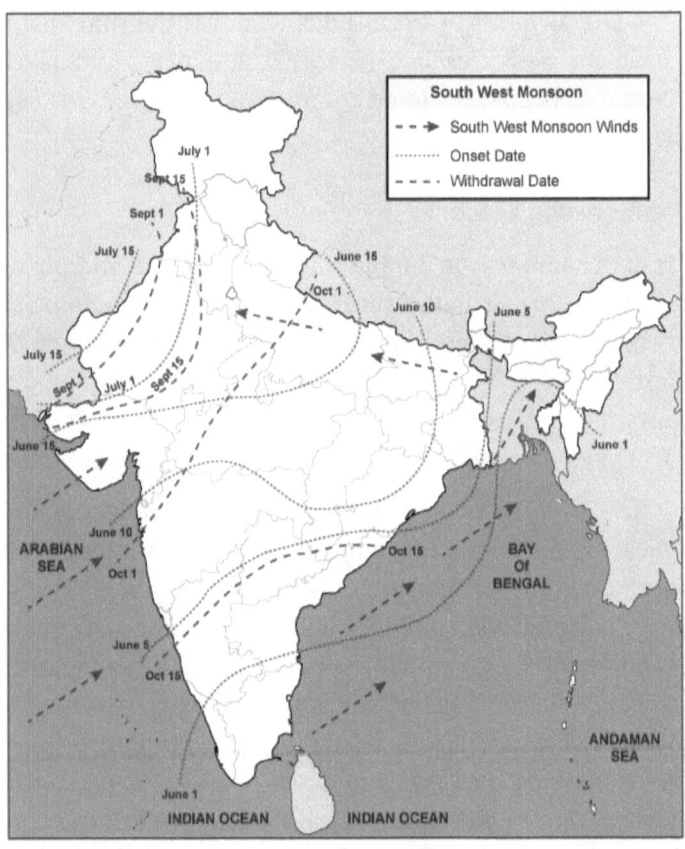

And in North-East India, it forms as a Tornado and sometimes as a waterspout in the river Brahmaputra. It is the only region in India, where Tornado forms and it causes lot of damage. In Bangladesh also Tornado forms near river Padma (Ganga and Brahmaputra joined together) and causes heavy damage to properties and human lives and in West Bengal, it is called as KalBaisahi. And in South India it is formed in Andhra, Karnataka and Tamil Nadu. In Tamil Nadu, it is mostly dominant in Salem and Coimbatore and in Thunderstorms and hailstorm is felt lot of super-cool water falls as toe.

Monsoon is supposed to be very important season for Indian sub-continent. The Government of India consults meteorological department and the planning and development of the country is done. And as India's whole economy is based on Agriculture, the south west monsoon plays a very important role as it brings rain to almost the entire country, farming entirely depends on this and the country's drought and the flood situation depends on that the monsoon in the rainy season of the country starting from June to November. Monsoon winds mainly brings water from the Arabian sea to the main land.

The South West Monsoon winds hits exactly the south of Kerala on June 1st. Just fifteen days before the monsoon, the hot summer makes the country extremely dry and sun scorched the entire Indian sub-continent. The climate becomes so sultry, people crave for rain as rain brings coolness in the sub-continent. And the western, eastern and North India gets water.

The monsoon winds (south west) that hits the south Kerala coasts in the 1st week of June spreads throughout the western coast of India up to Maharashtra and Gujarat. The monstrous south-west monsoon winds which lot of water from the Arabian Sea are blocked by the high rise Western Ghat mountains, which results in heavy rain in the western coast of India (is Malabar coast) and this last till the month of September. And the Eastern coast of the Indian peninsula get only the dry winds. And this wind blow with great speed. And this is utilized in the production of electricity in the western districts of Tamil Nadu (i.e. Coimbatore, Madurai and Thiruchi) and Western districts of Karnataka like Darwar, Hubli, Davangere and Belgaum districts.

These dry monsoon wind reaches the sea of Bay of Bengal and carries water from the sea and again re-enters the peninsula from Northern Andhra Pradesh, Orissa and Bengal coasts and again spreads over entire North India plains is the Gangetic plain and heavy rain is experienced all over India.

This is the monsoon winds brings rain to most parts of India barring Tamil Nadu in the Southern India and Rajasthan, the other parts of India mainly the Indo-Gangetic plain in Northern India and the Western Konkan coast of Kerala, Karnataka and Maharashtra. It extends up to the foothills of Himalayas. And it reaches up to Ganga Sagar in Rajasthan. And where the monsoon does not reach in Thar dessert in Rajasthan.

Post monsoon

After the monsoon showers blown over all over the country till September last. The post monsoon winds set in the Bay of Bengal in the months of October, November and December.

The low-pressure area forms in the Bay of Bengal sea. And it develops into a deep and strong low-pressure area, and gathers lot of water from the Sea (the Bay of Bengal) with a strong eye in the centre a well-marked low pressure and strong winds flow towards the southern coast of the peninsula.

These winds floating from the severe low pressure area in the bay of Bengal are called as North-East monsoon or (post-monsoon) winds, they contain lot of moisture and bring rain to Tamil Nadu, Andhra coasts, when the low pressure area becomes strong with eye formed by the walled clouds (CB clouds) in the centre, it becomes very strong and strong winds smash the coast with a speed more fifty to sixty kms. These wind earns heavy devastation in the coastal area and the fisherman are warned by the CWC (Cyclone Warning Centre) of India meteorological not to venture into sea through weather bulletin forecasts in the All India Radio (AIR) or Doordarshan Kendra (DD).

When the low pressure is not well formed, It doesn't appear like a circle with clear formal district eye structure then it is called as the 'trough' or low pressure along the eastern coast of India. There is a plenty of rainfall as there are too many moisture bearing clouds along the south India coast line.

And this is the time the southern part of the peninsula experiences very heavy rainfall as the trough moves very slowly along the coast line e.g. The heavy rain and floods experienced in the 1st week of December 2015 in the city of Chennai where the entire city was cut off from the rest of the main land. But although floods costs heavy devastation and adverse conditions are experienced but still it is welcomed as it brings plenty of water for the whole year.

The cyclone formed in this season causes wide spread devastation in its path all over the southern peninsula. Eg. Ockhi, Vardha, Gaja. All experienced in South India.

CONCLUSION

Although, when flying in aircrafts to different places, the pilot and the air hostess take care of the passengers for a pleasant journey, there are lot of unsung heroes like Airtraffic controllers, meteorologists and technical experts and communicators who also work (slog) in airport day in and day out on ground, who toil out so that the passengers have a safe and comfortable journey. This book is dedicated to all those unsung heroes, who are given little less importance in Aviation. I proudly say that I belong to one of their categories.

www.ingramcontent.com/pod-product-compliance
Lightning Source LLC
Chambersburg PA
CBHW021023180526
45163CB00005B/2089